For Dad

Copyright © 2012, Scott L. Haynes
All rights reserved
ISBN 978-0-615-84631-6
This edition fifth printing, Sept. 2020
www.roadietheranchdog.bigcartel.com
This book or parts thereof may not be reproduced without permission.

My favorite time of day when anything is possible HIP-HIP-HOORAY!

And, say "good morning" to my friend, the bull

My drink is cut short
I'd better hurry up

I run like the wind
I'm a fast little pup

I leap in the air
a smile on my face

In less than a minute
we're out of the yard
over the bumpy cattleguard

My nose in the breeze
we're well on our way
I smell sagebrush, skunk
and, freshly cut hay

The wind blows my ears
bugs hit my face

I just swallowed a grasshopper what an interesting taste

Friendly neighbors
moving cows
wearing grins from ear to ear

I need to stay right where I'm at
I caused a wreck last year

The horses sure are crackin' up and, let me tell ya, folks

You would be too
if you heard
a meadowlark telling jokes

A daydreaming colt
jumping and kicking
playing on the hillside

Dreaming he's a world-class buckin' horse that no cowboy can ride

There's Farmer Fred
and his pretty wife, May
racing each other
on bales of hay

Fred and May
sure are neat
it's time to go to town
and grab a bite to eat

My pals are here
what a great day

We run around
like wild animals
to pass the time away

You never know what you'll see maybe licorice-loving antelope doing tricks for candy

Or, our rancher friend
who likes to hunt
he lets his cows ride up front

A skittish horse who's a bit cuckoo
is still no match
for cowgirl Sue

A couple cowboys, roping some steers
how this wreck happened
is very unclear

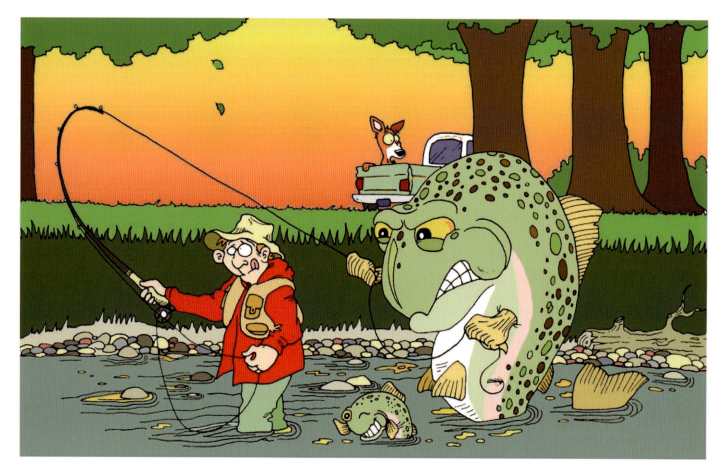

An evening fly fisherman with trouble on his line

An elk scratching his rear on a ponderosa pine

Another day done
what a treat it has been

I can't wait to wake up real early